ELECTRIC CARS

James Taylor

SHIRE PUBLICATIONS

Bloomsbury Publishing Plc

Kemp House, Chawley Park, Cumnor Hill, Oxford
OX2 9PH, UK

29 Earlsfort Terrace, Dublin 2, Ireland

1385 Broadway, 5th Floor, New York, NY 10018,
USA

E-mail: shire@bloomsbury.com

www.shirebooks.co.uk

SHIRE is a trademark of Osprey Publishing Ltd

First published in Great Britain in 2022

ISBN: PB 978 1 78442 491 6
           eBook 978 1 78442 492 3
           ePDF 978 1 78442 494 7
           XML 978 1 78442 493 0

22 23 24 25 26   10 9 8 7 6 5 4 3 2 1

Typeset by PDQ Digital Media Solutions, Bungay, UK

Printed and bound in India by Replika Press Private Ltd.

COVER IMAGE

Front cover: An Electric Smart car on charge (Alamy).
Back cover: The British road sign indicating a charging
point (Crown Copyright/Open Government Licence).

TITLE PAGE IMAGE

The CitiCar was designed and produced in Florida
from 1974 to 1977. (Eric Kilby/CC-BY-SA-2.0)

CONTENTS PAGE IMAGE

The dashboard ot the 2020 Honda E Advance, which
deliberately tackled the urban runabout market,
offering a range of just 125 miles but driving dynamics
that tempted the driver to use it for longer runs.
(Honda Wales/CC-BY-SA-4.0)

CONTENTS

Baker Electric Vehicles

The Aristocrats of Motordom

The Baker "Queen Victoria"

Baker Electrics are safest to drive—easiest to control—simplest in construc and have greater speed and mileage than any other electrics. Where quality efficiency are desired Baker Electrics are invariably the choice of discriminating and women who want elegant appointments combined with mechanical perfec

A request will bring to you our complete catalogue of Baker Electric Runabouts, Coupés, Roadsters, Landaulets, Broughams, etc.

THE BAKER MOTOR VEHICLE COMPANY,   33 W. 80TH STREET, CLEVELAND, C

Agencies in all Principal Cities.

EARLY DAYS

A T A TIME when electric or hybrid electric cars appear to provide the future of personal road transport, it is no surprise to find car manufacturers dredging up claims to early experiments with battery propulsion. In fact, the earliest experiments with battery-powered cars pre-dated the founding of any of today's car manufacturers. They were, however, just experiments.

The defining feature of an electric car is its use of electricity to power the wheels that provide motion. That electricity must be stored on board the vehicle, and as a result, electric cars are and always have been entirely dependent on the storage capacity of their batteries and on the ability of these batteries to be recharged.

The first true battery was invented by Alessandro Volta in 1800, but it suffered from numerous problems and was impractical except as an experimental tool. The technique of storing electricity in a battery was gradually developed by several inventors over the next half century, but the batteries of the period could be used only once: their usefulness ended when the chemical reactions that created electricity were spent.

None of that prevented inventors from seizing on this new source of portable power and designing battery-powered vehicles in the first half of the nineteenth century. Working independently of one another, individuals in Hungary (1828), the Netherlands (1835), the USA (1835), and Scotland (1839) came up with experimental vehicles, but the details and precise

OPPOSITE
US makers promoted electric cars as easy to drive and therefore suitable for women; it was an attitude very much characteristic of the times. This advertisement from 1909 is for the Baker 'Queen Victoria' electric car.

Gustave Trouvé added batteries and an electric motor to a British-made tricycle in 1881.

dates of some of these inventions are obscure. As demonstrations of theory, they were a start, but they were not by any means practical passenger-carrying vehicles.

The invention of the lead–acid battery by the French physicist Gaston Planté in 1859 brought a key advance in battery technology: it could be recharged by passing a reverse current through it. However, even then it would be some years before such batteries would become readily available. It was only after some major improvements to the design of the lead–acid battery by another French scientist, Camille Alphonse Fauré, that mass production of batteries was made feasible.

A number of experiments with battery-powered road vehicles followed in Europe. Quickest off the mark was the French inventor Gustave Trouvé, who adapted a German Siemens electric motor to power a British-made Starley tricycle, and tested it on a Paris street in April 1881. Developing his ideas further, he presented an electric car at the Exposition internationale d'Électricité in Paris that November. This unfortunately proved to be a dead end as he was unable to patent his ideas, so he turned his attention towards electricity for marine propulsion instead.

Not long afterwards, the British inventor Thomas Parker began to look at the viability of an electric-powered car for his company, Elwell-Parker of Wolverhampton, which was established in 1882 as a maker of accumulators and would later be responsible for major urban electrification projects such as the London Underground and the tram systems in

In 1887, JK Starley in Britain built an electric version of his company's tricycle – the latest version of the model that Gustave Trouvé had electrified six years earlier.

Birmingham and Liverpool. Parker had an experimental electric car running by 1884, and his experiments continued after the company became part of the Electric Construction Corporation in 1889. Notable among them was an electric dog cart that was built in 1896. Meanwhile, in Germany, the inventor Andreas Flocken produced his idea of an electric car in 1888.

In Germany, Andreas Flocken built an electric car in 1888. This is a modern replica of the original.

In truth, none of these early electric vehicles was yet a viable car. Small quantities were built for sale, but those who could afford them would certainly have had a horse-drawn vehicle for everyday use, and would have treated their electric car as a weekend indulgence. Petrol-powered cars were also making an appearance in this period, as were steam-powered cars, and would have been treated

in exactly the same way. At this stage, it was by no means clear which type of propulsion would predominate.

Nevertheless, public interest in self-propelled vehicles increased greatly towards the end of the nineteenth century. From 1897, a fleet of battery-powered taxicabs took to the streets in London. Designed by Walter Bersey, they made a distinctive humming noise when in motion and soon gained the nickname of Hummingbirds. Unfortunately, the weight of the batteries caused both excessive tyre wear and vibration that in turn damaged them, and the Hummingbirds were withdrawn after a couple of years.

The original *La Jamais Contente* of 1899 no longer survives, but this replica faithfully recalls its striking torpedo shape.

A small fleet of electric hansom cabs also began operation in New York in 1897, and a telling comment in *The Engineer* magazine for 7 July 1899 notes that the Paris Fire Brigade had chosen a new electric-powered fire engine because 'petroleum motors, despite the considerable advance that has been made of late years in adapting them to vehicles, cannot always be relied upon.' Also in Paris at that time, an interest in electric carriages was developing among the wealthy. At the first Paris Motor Show in 1898, ten different companies displayed no fewer than twenty-nine electric vehicles.

Two companies, called Jeantaud and Jenatzy, vied with one another for the Parisian market, and each indulged in publicity stunts designed to demonstrate its superiority. The speed of their products was key, and a Jeantaud vehicle reached 92.78 kilometres per hour (57.65mph) in March 1899. Determined to win, Camille Jenatzy designed an electric car to seize the record, and at the end of April (the exact date is disputed) his torpedo-shaped, light alloy electric car broke the 100km/h (62mph) barrier and reached a speed of 105.88km/h (65.79mph). Named *La Jamais Contente* (which translates roughly as 'The Restless One'), the car depended on batteries that powered an electric motor driving each rear wheel.

There were other interesting developments in Germany. Ferdinand Porsche, who later became one of the country's leading motor car engineers, developed an electric car capable of 25km/h while working for an electrical equipment company in 1898. The following year, he went with his invention to the Lohner carriage company in Vienna, which wanted to develop an electric powertrain for coaches, and the first experimental Lohner–Porsche Electromobile won a 50-kilometre race held as part of the Berlin Motor Show. Its twin 2.5PS electric motors gave it a top speed of 37km/h (23mph).

The European press got hold of the story and Lohner received its first order for an electric car from EW Hart, a British coachbuilder in Luton. However, Hart specified that his car should be capable of running on petrol as well as electricity, should have drive to all four wheels, and should have room for four passengers. Lohner and Porsche called the vehicle that resulted *La Toujours Contente* ('The Permanently Satisfied One') in a sly dig at Jenatzy's 1899 record-breaker. With a 14PS electric motor at each wheel, this was a huge machine that carried

OPPOSITE
In London, the Bersey Electric Cabs looked like the way forward, but the experiment was ultimately a failure.

La Jamais Contente had an electric motor at each wheel, with a chain drive to the axle. This is again the replica.

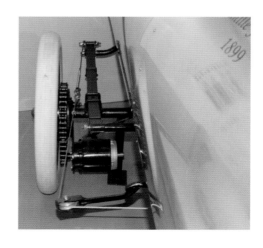

The pioneering Lohner–Porsche hybrid was an ungainly looking machine, but the technology worked well enough to lead on to limited production. This one dates from 1902.

The Porsche Museum in Stuttgart, Germany, has this display cutaway of the electric motor installed at each wheel of the Lohner–Porsche car.

1.8 tonnes of batteries and weighed more than 4 tonnes overall. It ran in several competitions and displays over the next few years but was far too expensive for popular consumption. So Lohner adapted the new drivetrain technology for large commercial vehicles, and by 1906 had sold more than 300 such machines. Porsche, meanwhile, pursued the technology with various racing machines before moving to Daimler–Benz as chief engineer in 1906.

Electric power was still just one option for propelling a passenger-carrying vehicle, and the electric car was in competition with both petrol-powered types and steam-powered types. It did offer certain advantages over petrol types, as it was entirely free of their smells, noise and vibration, and it did not require either hand-cranking to start the engine or a mastery of tricky gear changes. As compared with a steam car, its main advantage was that it did not require a long start-up time,

There was considerable enthusiasm for electric cars in the USA in the first decade of the twentieth century. This 1907 GM Electric looks very much like contemporary steam and petrol models.

as there was no need to wait for the water in the boiler to be heated from cold to deliver the necessary steam pressure.

In the USA, electric cars were catching on fast. In 1899 and 1900, they outsold all other types, and the American Census reveals that 28 per cent of all 4,192 cars made in the USA in 1900 were electric. Shrewd marketing soon saw electric cars being touted as ideal for women drivers because they were clean and fuss-free. As long as they were used only for short trips around town, to visit friends and go shopping, they worked well enough. At that stage, the electricity distribution network was still limited, and electric cars had to be charged at dedicated charging stations. However, over the next decade the spread of electric lighting made it possible to charge an electric car simply by plugging it in to a domestic socket.

The peak year for electric car sales in the USA was 1912, but by this time there had been two key developments in petrol-powered cars. One was that Henry Ford's use of mass production for the Model T he introduced in 1908 had brought prices right down; by 1912, an electric car might cost three times as much as a Model T. The other was the

OVERLEAF
The ability to recharge a car from a domestic supply was welcomed as electricity spread across the USA. This 1919 picture shows a Detroit Electric model on charge.

introduction that year of electric starting by Cadillac, which removed the inconvenience of hand-cranking.

Yet even Henry Ford kept a wary eye on the popularity of electric cars. In the 1890s he had worked for electrical pioneer Thomas Edison, and by 1913 the two men were collaborating on the design of an electric car. During 1914, a second prototype was based on a Model T chassis, and in January that year, Ford told *The New York Times*:

> Mr Edison and I have been working for some years on an electric automobile which would be cheap and practicable. Cars have been built for experimental purposes, and we are satisfied now that the way is clear to success. The problem so far has been to build a storage battery of light weight which would operate for long distances without recharging. Mr Edison has been experimenting with such a battery for some time.

The project fell by the wayside when Edison's nickel–iron batteries were discovered to have too much internal resistance to do the job. When lead–acid batteries were tried instead, their extra weight quickly put paid to the project. It was a problem that would dog the electric car for many more years.

By this stage, Europe was about to be plunged into war. The horse remained the predominant choice of transport, and the short range and low top speed of battery-powered vehicles debarred them from serious consideration by the military. Petrol power was selected for motorised transport, and was the only viable option for aviation. Meanwhile, in the USA, the popularity of the petrol-powered car was increasing, and further new discoveries of oil made abundantly clear that it had a long-term future. Five years after the war's end, records show that just 391 electric vehicles were manufactured in the USA, as compared with 3.18 million petrol types. Electric power had lost the battle everywhere.

INSUFFICIENT INTEREST

For the best part of forty years, electric cars fell right out of favour. Electricity was still favoured as a source of power for transport, but the electric trains, trams and trolleybuses that were a feature of the inter-war years and beyond depended on hugely expensive infrastructures. There was a recognition that they could not carry sufficient power in onboard storage batteries, and that they must therefore pick up electricity from a permanent supply as they were moving. Trains used an electrified 'third rail' or an overhead supply; trams used an overhead supply to avoid the danger of a 'live' rail in city streets. The trolleybus improved flexibility and reduced the cost of the infrastructure because it needed only an overhead supply and no rails. Trolleybuses did have onboard batteries for low-speed manoeuvring at their maintenance depots, but there was never any question of using these for longer distances.

Battery power was not forgotten altogether, of course. The battery-powered delivery vehicle had existed before the First World War and was still considered a cost-effective solution for urban areas where high speeds were not required and distances were short. The batteries of an electric delivery vehicle could be charged overnight and would hold enough power to see it through the following day's work. The popularity of such specialist vehicles ensured that the technology of the short-range battery-powered vehicle was perfected, but there was no real incentive to develop it further. Battery power for vehicles remained a niche interest.

The electric trolleybus was welcomed in towns and cities where the local authority had its own electricity generating station. This London trolleybus dates from the early 1930s: note the two booms on the roof, which collected power from overhead wiring.

In Britain, the famous London department store of Harrods operated a fleet of electric delivery vans in the city after 1919. The first ones were American-made Walker types, but in 1933 the store started work on designing a 1-ton van of its own. A prototype was running by 1935, and Harrods built a fleet of sixty in its own workshops between then and 1939. The batteries were carried under the floor and gave a range of 60 miles on a single charge with a top speed of 18mph. The vans proved enduringly reliable. The oldest ones remained in service for well over thirty years and were a familiar and much-loved sight in the west end of the capital.

Even more familiar all over Britain were electric milk floats. Daily deliveries of fresh milk and other dairy products to households had once depended on horse-drawn vehicles, but the 1930s saw the rise of battery-powered vehicles and by the 1950s these had become the norm. Among the major manufacturers were Smith's Delivery Vehicles, which in 1949 was separated from its parent company, the commercial body

London store Harrods developed their own design of electric delivery van, which was ideal for short-distance work and remained in use until 1970.

builder Northern Coachbuilders. Smith's went on to become the world's largest maker of electric vehicles and contributed to several electric-vehicle programmes by other manufacturers, eventually being bought out by its own US subsidiary in 2011, but sadly ceased trading in 2017 as revenues and funding dried up.

Battery power did see a brief revival of interest in occupied France during the Second World War. The occupying German

The milk float was once a familiar sight in Britain, having enough battery power to see it through a day's deliveries and returning to the depot to be recharged overnight. These three, of various types and ages, are pictured at the headquarters of Collins Dairies in Southend in 1969.

forces requisitioned almost every serviceable vehicle, and those that remained could not be driven far because of a shortage of petrol, which was both rationed and very expensive. The result was a pressing need for transport to keep essential civilian services running, and during 1941 three designs of electric car entered production. Not one of them was ever very numerous, but they met a need.

One was a utilitarian two-seat city runabout designed and manufactured by Peugeot that was called the VLV (*Voiture Légère de Ville*, or 'Light City Car').

When needs must… the VLV was Peugeot's response to a need for transport when fuel supplies became scarce in occupied France during 1941.

Its appearance was almost comical, with a wide front track and twin rear wheels mounted to give the appearance of a three-wheeler. The VLV had four 12-volt batteries mounted under the bonnet that powered a Safi electric motor driving the rear wheels. Top speed was 22mph and the batteries gave up to 50 miles on a single charge.

The cars were built at the La Garenne factory between 1941 and 1945, and most

Looking more like an ordinary car, this was the CGE Tudor of 1941, another electric car created in response to the difficulty of obtaining petrol in wartime France.

of the 377 examples built were used by postal workers and doctors.

The second was a collaborative venture between car maker Jean-Albert Grégoire and the Compagnie Générale d'Électricité (CGE). Known as the CGE Tudor, this was a more realistic-looking and larger four-wheel cabriolet with its electric motor mounted amidships. The sixteen batteries were carried in two

groups under the bonnet and behind the seats, and accounted for nearly half the total weight of the car; they were recharged by a regenerative braking system. With a maximum speed of 36mph and a range of 55 miles on a single charge, the CGE Tudor was available only to privileged members of French society such as industrialists and civil servants, and just 200 examples were built.

The third came from aircraft maker Bréguet Aviation, which was granted permission to produce an electric car to keep its workforce occupied, aircraft production having of course been suspended under the German occupation. The

Again, weird in appearance, but meeting an urgent need as rapidly and economically as possible, this was France's Bréguet A2 electric car. This 1942 model is now a museum piece.

Electric delivery vans remained common for many years. This photo shows a line-up of brand-new electric postal service vans in East Germany in 1953.

The BMA Hazelcar was a serious early 1950s attempt in Britain to create an electric car, but only a handful were built, the last one in 1955.

Bréguet A2 was drawn up with a simple two-seater body in which ease of manufacture was probably a higher priority than the appearance, which had something of the streamlined aircraft about it. Its long tail contained a two-stage Paris-Rhône electric motor and some of the six 12-volt batteries. Like the Peugeot VLV, the Bréguet A2 was a four-wheeler with a wide front track and narrow rear track, and in this case the first (36-volt) stage of the motor gave low speeds while the second (72-volt) stage gave higher speeds. The realistic range on a single charge was 40 miles, but the car's only real advantage was its availability. Once again, about 200 were made.

With the return of peacetime conditions in the later 1940s and the resumption of civilian petrol supplies, there was once again no imperative to manufacture electric cars. There were sporadic attempts, and in the late 1950s two of the more interesting ones were made in the USA. Both were prompted by commercial interests rather than by customer demand, and both failed.

It was in 1959 that car maker AMC (American Motors Corporation) and battery maker Sonotone announced plans for an electric car with a 'self-charging' battery. Sonotone at the time had developed nickel–cadmium batteries that were lighter than traditional lead–acid types and could be recharged more quickly, and this project would have promoted their new technology, but proved abortive. More fruitful was the Henney Kilowatt, which was also announced in 1959.

The impetus for this one came from the National Union Electric Company, the US makers of Exide batteries, and the car was built in collaboration with the established custom body builder Henney Coachworks. The Kilowatt used the body of the French Renault Dauphine, and the first cars had a General Electric motor powered by eighteen 2-volt batteries. Their top speed of 40mph and 40-mile range were inadequate, and the car was redesigned for 1960 with twelve 6-volt batteries, giving double the power capacity. The car's makers had bought 100 Dauphine body shells for initial production, but the customers could not be found and probably only 47 Kilowatts were ever completed. Most went to electric utility companies, and fewer than fifteen found private buyers. High prices and insufficient interest killed the project in 1961.

It looks like a Renault Dauphine, because it was built into the shell of one. This is the 1959 Henney Kilowatt that entered low-volume production in the USA.

RESURGENCE OF INTEREST

THE 1960S BROUGHT the first signs of concern about rapidly increasing car use. In the cities of the developed world, road congestion was becoming a problem, and in the US state of California there were serious concerns about the health hazard from polluted air in the major cities, to which cars were considered a major contributor. Out of this grew the idea of a small city car that would reduce congestion by taking up less road space, and would reduce pollution by using less fuel. It was a small step from there to the idea of a small city car powered by a non-polluting battery that could be recharged from a domestic electricity supply.

The idea of a small two-seater with a limited range and limited speed quickly became established as the norm. These limits were really imposed by the battery technology of the time, but were not seen as problematic for a city runabout. Careful choice of materials would enable lightweight construction, and this would further reduce the drain on battery power. So the 1960s saw some of the earliest concepts for a battery-powered city car.

In Britain, an early one came from Scottish Aviation, an aircraft maker in Prestwick. By mid-1965, the company had developed a battery-powered two-seater they called the Scamp, and began negotiating with the CEGB (responsible for British electricity production) about selling it through their regional showrooms. After a successful demonstration of the prototype, a further dozen test cars were built in

OPPOSITE
Designed and made in Florida, the Citicar was an American response to the Oil Crisis of the early 1970s. It entered production and became relatively common, but the quirky design won it few friends.

1967, but battery life proved disappointing, and a thorough durability test showed up serious suspension weaknesses, so the project was abandoned. Nevertheless, the idea had gained traction, and in 1966 the Electricity Council (which oversaw electricity supplies in England and Wales) announced a competition to produce a viable electric car. Funded by a Greek shipping millionaire, the winner was the Enfield 8000, developed and built on the Isle of Wight and viable by 1969. The Enfield's 8bhp electric motor drew power from eight 6-volt batteries that could be recharged from a domestic socket, and on a single charge the car could run for up to 56 miles. Production began in 1973 on the Greek island of Syros (although actual assembly remained on the Isle of Wight), but sales were slow, partly because of

In the 1960s, electric cars were more or less universally thought of as dumpy little two-seaters. The Scottish Aviation Scamp of 1965 was typical; this one is a preserved 1967 example.

A carefully positioned female model prevents the Enfield from looking as short as it really was. This brave try did at least enter production, but cost killed it.

the limited range and partly because of the cost: an Enfield cost roughly twice as much as a Mini. Production was halted in 1976 after 120 had been built; of those, 65 were used by the Electricity Council itself and by electricity boards in the south of England.

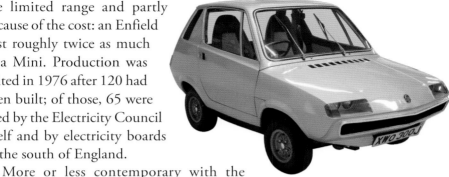

More or less contemporary with the Enfield was a model from car maker British Leyland, whose Morrison Electricar milk float division made it, using Crompton electrical equipment. It used several standard parts from the Mini to keep its manufacturing cost down, and was displayed at the 1972 Geneva Motor Show as a demonstration of corporate ability. With a GRP body styled in Italy by regular BL consultant Michelotti, the Electricar needed no fewer than twenty-four standard lead–acid batteries to power its two 3.9hp electric motors. It could be recharged from a domestic supply, would have cost about a penny a mile to run at 1972 prices, could reach 33mph and had a maximum range of 40 miles. But those batteries made

Italian designer Michelotti did his best with the Electricar from British Leyland, but the short two-seater package did not leave him much room for manoeuvre.

Even the Americans did not think big. The 1967 AMC Amitron might have added style into the equation, but it was still essentially a two-seater with rather odd looks.

General Motors showed their Electrovair II to members of the US Senate. The huge bank of batteries in the front was matched by a similarly sized pack in the rear, which between them gave an 80-mile range.

it heavy – some 30 per cent heavier than a standard Mini. It was a brave statement, but no production followed.

Meanwhile, the problem of air pollution from car exhausts was exercising the American Congress, and in 1966 it amended the 1963 Clean Air Act and passed bills that were collectively known as the Electric Vehicle Development Act. The legislation provided funding for electric car research, and Ford, Chrysler, General Motors and AMC all initiated electric-vehicle research programmes in response. By 1967, AMC had its Amitron concept, a city car with two interlinked battery systems. Lithium–nickel fluoride batteries were the primary power source, fed by fast-charging nickel–cadmium batteries that were recharged by a regenerative system, where the motor doubled as a generator during deceleration. The Amitron had a range of 150 miles at 50mph, and AMC hoped to offer production versions within five years as urban commuting and shopping cars. Sadly, the huge cost of the batteries caused research to be suspended. Although the Amitron prototype was recycled in 1977 under the name of Electron and was touted as a future product, it went no further.

While the city runabout was the main thrust of electric-car development in this period, there were also attempts to create an electric car that looked more like a conventional model. It was clear from early on that buyers would take more interest in a full-sized family car powered by electricity, and in the mid-1960s General Motors developed prototypes based on its rear-engined Corvair. The Corvair was the lightest GM model then in production; the first Electrovair appeared in 1964, and an improved one in 1966 was based on the second-generation Corvair. Weighing 1,000lb more than the standard petrol car, the Electrovair II had 532 volts of silver-oxide batteries in the front, chosen for their high energy density. These drove through a 115hp induction motor where the engine had been. GM claimed an 80mph top speed, acceleration from standstill to 60mph in 16 seconds, and a range of between 40 and 80 miles depending on the type of usage. But the Electrovair remained experimental: the batteries were very expensive, heavy, and wore out after 100 recharges.

In Germany, BMW also began to look at battery power in the late 1960s, focusing on an electric version of their

Sensibly, BMW experimented with a battery-powered version of an existing and much-liked car. Without the decals on its doors, this one was indistinguishable from the 1602 on which it was based.

compact 02 model saloons. A small fleet of test cars each carried a dozen lead–acid batteries (weighing a formidable 350kg) under the bonnet, and drove through a 43bhp Bosch electric motor mounted where the standard car's gearbox went. The rear wheels were driven through reduction gearing and a propshaft. A simple forward or reverse selector replaced the gear lever.

The BMW 1602e could reach a promising 62mph but at half that speed had a range of only 19 miles despite a regenerative braking system, where the motor doubled as a generator to recharge the batteries during deceleration. Two examples from the test fleet were presented at the 1972 Olympics in BMW's home city of Munich, where they featured in the opening ceremony and as stewards' transport for the marathon and other long-distance walking events. But BMW knew by then it was all for show: battery technology was once again the limiting factor.

Interest in electric cars received a shot in the arm after the fuel price rises that followed the 1973 Oil Crisis, and an early

The Oil Crisis prompted a further look at electric power, and the Italian Zagato company put its Zele into production in 1974. Yet with all their acknowledged styling skills, they were unable to move beyond the idea of a quirky little two-seater runabout.

'We regard this as a step in our programme to develop a commercially practical electric car,' said Ford's British branch of their 1967 Comuta car. However, only two were ever built.

design was the 1974 Zagato Zele, made by the manufacturing division of the Italian coachbuilder. Sold in some countries, including the USA, as the Zagato Elcar, this reached about 500 examples before production ended in 1976. Even more numerous was the CitiCar that was made in Sebring, Florida between 1974 and 1982, latterly in revised form as the Comuta-Car. A claimed 4,444 were built – a tiny number in a big country like the USA, of course. Both of these were two-seat city runabouts with a range of about 50 miles and neither really advanced the technology, but the public had now become more receptive to what they offered.

GM picked up their earlier experiments again in 1976, this time determined to develop an electric version of their latest sub-compact, the Chevrolet Chevette hatchback. The aim was to be prepared for the further rise in fuel prices that was expected to come, but the Electrovette was compromised from the start. First, it was designed as a second car for local use only, and second, it was shortened to reduce weight and counter the extra weight of the batteries. When the first trials with nickel–zinc batteries were disappointing, the engineers fitted cheaper lead–acid types. Powered by a 63hp motor, the car had a top speed of 53mph and a claimed range of 50 miles at 30mph. Realistically, not much progress had been made since the 1960s, and GM admitted that new battery technology was needed before they could build an affordable, consumer-friendly electric vehicle in quantity.

By the end of the 1970s, despite considerable efforts in several countries, the electric car was still fundamentally impractical. It was too heavy, thanks to the enormously heavy battery pack it had to carry; it was generally slow by comparison with conventional petrol-powered cars; and its range was far too small. Most prototypes were distinctly lacking in glamour, and too many made up for that by offering a kind of apologetically quirky or toy-like design.

Nor did electric cars yet offer what buyers really wanted. Used to being able to carry at least four people in a car, and to being able to drive as far as they wanted subject to the availability of fuel, they found the two-seat configuration and limited range unsatisfactory. For most people at this stage, an electric city car could only be an addition to a normal petrol car in the garage – which gave it the status of a rather expensive toy.

The German answer was once again more pragmatic, and Volkswagen announced an Elektro-Golf in 1976 alongside the original Golf GTi. At least it looked like a real car, but it did not progress beyond the experimental stage.

MARKING TIME, 1980–99

ELECTRIC CARS WERE still very much experimental during the 1980s, in both Europe and the USA. After the initial surge of interest that followed the 1973 Oil Crisis, there seemed to be no imperative for the car makers to develop viable electric cars, except perhaps as a demonstration of their own capability. There was no real customer interest in alternative fuels of any kind as oil prices stabilised, and the industry began to focus more widely on diesel cars as a means of reducing the cost of fuelling personal transport. Up to this time, every electric car that had been developed had also reduced convenience and mobility; diesel cars did not suffer from that disadvantage.

In this decade, therefore, there were no real strides in electric-car technology. Nevertheless, several major manufacturers kept their electric-car programmes ticking over. This was partly to gain publicity, and partly to make sure they would not be caught out if there was a sudden upsurge in customer interest in electric cars. Other alternative-fuel programmes were also kept ticking over, one being the development of hydrogen-powered vehicles. Yet cost and technological barriers made sure these offered no clear advantages over electric power.

There were two important turning-points in this period. One was the establishment in 1988 of an Intergovernmental Panel on Climate Change that would take the lead in highlighting the contribution made by vehicle exhausts to the issue of global warming. The other came in 1990, when the

The first of Volkswagen's CitySTROMer cars was based on a Golf Mk I in 1981. About a hundred were built, but restricted range was still a problem.

California Air Resources Board exercised its unique right to set state-specific emissions standards independently of the Federal Government by passing a zero-emissions mandate. This optimistically called for 2 per cent of cars sold in California to be emissions-free by 1998, and it provided a reason for car makers in the USA and in Europe to renew their efforts to develop viable electric models. None of those manufacturers wanted to be in a position where it could no longer sell cars in California, the third-largest and also the wealthiest state in the USA. When a consortium of manufacturers later won a legal challenge to the Californian legislation, it took some of the impetus out of electric-car research, but some important steps had already been made.

Fairly typical of the electric-car programmes of the 1980s was the one run by Volkswagen in Germany. The company had first looked at electric power in the 1970s in the wake of the Oil Crisis, and in 1976 had announced its development of the Elektro-Golf. But the development programme stalled, and during the 1980s seemed largely to be a flag-waving exercise. In 1981 it delivered the first Golf CitySTROMer (*Strom* is German for 'electric current'); in 1985 came a CitySTROMer Mk II based on the latest revised Golf (but bringing no other important improvements); and then a model based on the booted Jetta version of the Golf did demonstrate some improvements. About a hundred examples of the Mk I CitySTROMer were built, but their range was a feeble 37 miles on a single charge. There were around 70 experimental versions of the Mk II. The Jetta-based version had a better range, offering 118 miles with lead–acid batteries or 155 miles with the later sodium–sulphur batteries. But the technology was still not sufficient to make the electric Volkswagens good enough to become mainstream products.

In France, Renault and the Peugeot-Citroën combine followed a slightly different path, focussing initially on light vans and persuading government departments and government-owned organisations to take test fleets. Like Volkswagen, they did not develop all-new models but adapted

Plain common sense: BMW's E1 used the latest battery technology (which proved less than stable) and provided four seats in a most attractive shape. One early prototype burned out after a battery fire.

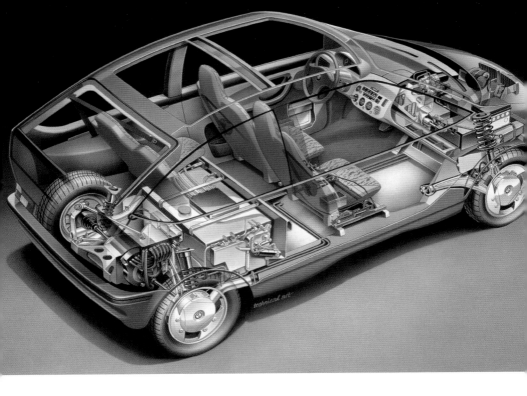

Rather than squeezing an electric drivetrain into an existing model, BMW designed the E1 around its electric components. All the vital powertrain elements were neatly packaged under the floor at the rear.

existing production types to take electric power; first came the Renault Express van in 1985, and then in 1989 came the Peugeot J5 van, followed soon after by the closely related Citroën C25 and C15 types.

The 1990s would be different. First, there was the impetus of that Californian legislation, and second, there were developments in battery technology. Sodium–sulphur batteries looked like providing an answer for a time – although it became clear that they could overheat and lead to vehicle fires – and then the availability of nickel–cadmium batteries from around 1992 provided the next step forward. These were lighter than earlier types, offered greater energy density, could be recharged more quickly, and lasted longer before needing replacement. Yet they were still not quite the answer.

One notable concept car that appeared in 1991 was BMW's E1 city car. Its development was prompted by that

1990 legislation in California, and it was typical of the times that BMW Technik (the company's advanced vehicles think-tank) should conceive of it as a city car. Despite a long-running electric-vehicles programme, they believed that a full-sized family car was not yet viable.

The E1 depended on a sodium–sulphur battery driving a motor mounted on the rear axle. The arrangement gave a maximum range of 155 miles – excellent for the time – and the battery could be fully recharged overnight from a domestic power supply. As a demonstration of the possible, it was very impressive, and BMW had made the car particularly attractive too, with a lightweight body made of aluminium and plastic. However, although a Mk2 model at the 1993 Frankfurt Motor Show took the project a stage further by combining the 45bhp electric motor with an 82bhp petrol engine (from the K1000 motorcycle), the project was

The E1's plug looks crude by modern standards, but BMW had thought carefully about where to stow it – behind a hinged section of their traditional twin-kidney grille.

There was no question that BMW had designed a car with real appeal as well as a remarkable range for the time. Unfortunately, the cost of development was huge, and the company needed to spend its money on other projects with a more clearly defined future.

dropped. Cost was a major reason; this was an expensive programme and BMW needed the money more urgently for other projects.

Among those was its participation in the most important large-scale European experiment with electric cars of the decade. This involved all the leading German car makers, and was supported by the Federal Ministry of Education and Research (BMBF). Its aim was for Germany to take the lead in developing the technology to meet the Californian zero-emissions target. The German government was not widely supportive; its stance was that battery-powered cars were of no value within Germany because the country's power stations were all coal-fired and the extra demand for electricity generation would actually increase overall emissions. Nevertheless, the BMBF pursued the issue and arranged a long-term trial to discover if electric cars could meet the demands of everyday use.

The trial was carried out on the island of Rügen, in the Baltic Sea off the coast of Germany. It was chosen partly because its electricity was already provided by wind turbines, so the additional demand for charging the cars would not contribute more pollution. Another key factor was that Rügen had been a part of East Germany, and that government grants were available to assist technological integration with the West.

The experiment ran from 1992 to 1996. BMW, Mercedes, Opel and Volkswagen all provided electric versions of their current models, and these were given to 58 inhabitants of the island to use in place of their existing cars. BMW based their 'e-mobil' on a 3 Series coupé, Mercedes offered a 190E Elektro converted from their current small saloon, Opel provided converted Kadett estates and Volkswagen delivered their Golf

The Mercedes-Benz 190E Elektro was used in real-world user trials on the island of Rügen, but mainly proved that the technology was not yet fully mature.

The electric motor of the 190E Elektro was mounted almost conventionally at the front of the car. As the Rügen trials progressed, all the participating manufacturers did their best to improve the test cars.

CitySTROMer III, as well as some electric versions of their T3 Transporter.

Reliability was not a particular issue, although the inhabitants of Rügen quickly discovered the disadvantages of electric power at that stage of its development. Hills presented problems, batteries often ran out of charge at critical moments, and performance was poor. The four car makers made several changes to their electric-vehicle technology during the experiment, particularly to battery types and electric motors, and in 1994 they got together with battery makers and electricity utilities to present a report to the German government that demonstrated how half a million electric vehicles could be on the road by 2000. However, that report was largely ignored; the Rügen experiment was considered a failure; and the car makers themselves were investing too heavily in new diesel technology to pursue the electric option independently.

In France, Renault promised an electric version of its Clio hatchback as early as 1990, but it would be five years before it reached the market. Rival maker Peugeot Citroën meanwhile determined to take leadership of the electric-vehicle market, and developed electric versions of its Peugeot 106 and Citroën Saxo twins. These and the Elektro-Clio all reached the market in

Volkswagen's cars for the Rügen trial were Golf CitySTROMer III types, based on the conventional Mk III Golf. This was the under-bonnet view.

The Peugeot 106 Electric looked exactly like a standard 106, and that was the point. Unfortunately, less than 70 miles on a single charge was not enough, even if costs were reduced by renting batteries.

1995, and both companies adopted the tactic of selling the cars to the public but renting out the battery packs. The high rental cost proved a major deterrent, as did the mediocre performance, but Renault did manage to sell about 1,000 Elektro-Clios (mainly in Sweden) and Peugeot Citroën claimed about 12,000 of their cars before the scheme ended in 2003. Their range of 68 miles per charge was quite good by contemporary standards but clearly illustrated the continuing limitations of the electric car.

A display drivetrain of the Citroën Saxo at the Autovision Museum in Altlußheim, Germany, makes clear how much room the various components took at that stage of development.

In the USA, meanwhile, there was a slightly different approach. Controlled field trials were again on the agenda, but the typical American pattern was to build a fleet of test cars and lease them to users. When the experiment was considered finished, the vehicles would then be repossessed – although in some cases a few examples remained in private hands. Dodge went down this route with the TEVan (an MPV or minivan) in 1993, and GM followed with the Chevrolet S-10 pick-up in 1997–98. Ford then produced their Th!nk model in 1999, but none of these trials really took matters much further forward because ranges were still very limited. Japanese makers followed a similar pattern with electric cars in the USA from 1977; Honda leased examples of its EV Plus, Nissan of its Altra, and Toyota of its RAV4 EV. The latter was perhaps the most successful, with a claimed maximum range of 120 miles.

The most important of these trials was carried out by General Motors with its EV1 model. The company's good-looking 1990 Impact electric concept car had been favourably received, and GM's Advanced Technology Vehicles group was given the task of investigating the viability of such a vehicle. The EV1 duly appeared in 1996, drawing heavily on the Impact and in particular on its sleek looks, which in turn were inspired by GM's Saturn brand. It had a transversely mounted 137bhp electric motor driving the front wheels through a single reduction gear that was integral with the motor and differential.

By the time of the Peugeot 106 and Citroën Saxo twins, nickel–cadmium batteries offered the best option for electric cars.

In initial '1997 model' form the EV1 coupés used lead–acid batteries; 660 cars were built and were leased at

Renault developed the Zoom urban runabout with its Matra subsidiary and showed it in 1992. The rear wheels, extended here, folded forwards to create a shorter wheelbase for urban parking, and there was a 61bhp electric motor.

appropriate market rates to users in selected western US cities. The user base was later widened, and 457 examples of a second-generation or '1999 model' EV1 were built. These had lighter and more energy-efficient nickel–metal hydride batteries that

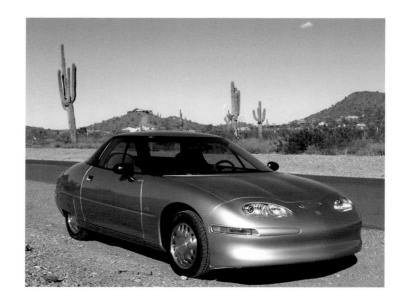

With the EV1, General Motors managed to combine an attractive and quite futuristic shape with an electric drivetrain that worked well.

The EV1's 137bhp motor drove the front wheels through a single reduction gear.

nearly doubled the range to 142 miles. However, in 2002 GM terminated the EV1's 'real-world engineering evaluation' and repossessed all the cars. Most were scrapped, although a few were deactivated and presented to museums, and a single running example was presented to the Smithsonian Museum. Despite favourable reports from users, GM had concluded that the EV1 was not commercially viable and that there was no customer demand for such a car.

GM decided in 2003 that the long-term evaluation of the EV1 had proved an electric car was not yet commercially viable. All the test cars were recalled – much to the disappointment of many users – and here they are lined up at a facility in Burbank, California, prior to being scrapped.

THE HYBRID OPTION

CAR MANUFACTURERS SHOWED considerable hesitation about committing to the development of fully-electric models in the final years of the twentieth century. The development costs themselves were very high – GM in America is said to have spent $1 billion on the abortive EV1 programme of the 1990s – and there was undoubtedly a problem with customer resistance. While buyers who took a strong interest in environmental issues were strongly in favour of zero-emissions fully-electric vehicles and were prepared to fund the extra purchase costs, there were many more potential buyers whose only interest was in saving costs. The fully-electric vehicle could not yet achieve that for them.

The major field trials of the 1990s also revealed a deterrent to electric-car use in both groups, which was the fear of running out of battery power while driving. This came to be known as 'range anxiety'. New battery technology was under development to address the issue, but it was not going to be available quickly or cheaply. If the electric car really did represent the future of motoring, and there were many influential people who believed that it did, there would have to be a more gradual transition to fully-electric vehicles.

The search for a zero-emissions solution was of course not confined exclusively to electric power. Some manufacturers had been experimenting for some time with hydrogen power as an alternative. This had its attractions but, like electric

power, it also presented some problems to which the solutions were likely to prove expensive. It was therefore by no means an obvious answer to a zero-emissions motoring future, and even those manufacturers who flirted with it also set up electric-vehicle development programmes.

The solution to a more gradual transition to electric power was an entirely pragmatic one, although it really came about almost accidentally as the result of continuing experiments with fully-electric vehicles. In business terms, hybrid electric cars would allow manufacturers to make progress with the development of fully-electric cars while providing an attractive half-way house to customers in the hope that the eventual transition to fully-electric power would be easier for both sides. So, from the late 1990s, several manufacturers committed development resources to hybrid electric vehicles, while at the same time not abandoning the idea of fully-electric power. Hybrids became very popular, but not at the expense of the

Land Rover in Britain produced this display concept called the Land_e in 2007, after receiving heavy criticism for the fuel consumption of its vehicles. The skeletal display unit, seen here at the Geneva Show, incorporated the technologies that the company planned to use in future hybrid electric vehicles.

fully-electric option; as the next chapter explains, there were some major strides forward in that area, too.

The essence of a hybrid electric vehicle is that it has a secondary onboard power source that can either assist the fully-electric mode or take over from it if needed. In practice, that secondary power source has always been the tried-and-tested internal combustion engine. There are multiple ways in which such a system can be configured, and in the early years of the twenty-first century a range of new terms was coined to describe them. As a warning, it is worth noting that some of these terms are often misused, but it is helpful to review what they are.

Among these terms, perhaps the most common are Parallel Hybrid, Range Extender, and PHEV. In a Parallel Hybrid vehicle, the internal combustion engine and the electric motor are designed to be used in tandem to drive the wheels; typically, the electric motor is used alone at low speeds, but the petrol (or diesel) engine cuts in to provide extra power for higher speeds and for acceleration. In a Range Extender type (also known as a Series Hybrid), one or more electric motors drive

By 2011, Land Rover had a viable hybrid powertrain that combined electric power with a diesel engine. This was one of its experimental Range Rover Sport PHEV vehicles, also known as Range_e types.

the wheels while the internal combustion engine is used when needed to power a generator that tops up the battery charge in order to provide continued mobility. The letters PHEV stand for Plug-in Hybrid Electric Vehicle, and the key characteristic of this is that the vehicle's batteries can be recharged from an external power supply – literally 'plugged in' to an electric main. Within these three categories, multiple configuration options allow a certain amount of overlap.

Two other terms that have become common deserve explanation, and these are Mild Hybrid and Full Hybrid. In a Mild Hybrid vehicle, the electric motor provides assistance to the internal combustion engine but cannot be used to power the vehicle on its own. With a Full Hybrid powertrain, either the electric or internal combustion sources can be used to power the vehicle, independently or in tandem.

The technological lead in hybrid electric vehicles came from Japan, although the basic concept was far from new; as the first chapter shows, Ferdinand Porsche had made it work by the end of the nineteenth century. However, car maker Toyota recognised that modern technology would make it infinitely more acceptable for everyday use. At the 1995 Tokyo Motor Show, the company displayed its Prius concept; the name is Latin and means 'first' – and indeed the Prius was the first viable modern hybrid electric car. The production version was introduced to the Japanese domestic market in 1997 and in 1998 won Japan's car of the year award; a slightly improved version was released for export markets in 2000.

The Prius was a straightforward everyday family car, with the unthreatening configuration of a five-door family hatchback. Its nickel–metal hydride battery pack was mounted under the rear floor and drove the wheels through a high-torque electric motor, and there was a 1.5-litre petrol engine as well, that powered an onboard generator when needed to boost the charge in the battery pack. The internal combustion and electric power sources were seamlessly integrated and required no special effort by the driver; the car's control system made all the important decisions and the Prius was completely autonomous. It did require the petrol tank to be topped up from time to time but its consumption of that fuel was minimal. The Prius set the global standard for hybrid electric cars, and for many years remained the global best-seller as well.

Not far behind were Honda, who were obliged to give their first hybrid a more radical appearance to help distinguish it from the Prius. So their Insight was released as a quite futuristic-looking two-door coupé, which probably took some of its inspiration from General Motors' EV1 electric car. It was previewed at the 1997 Tokyo Motor Show as the J-VX concept car and brought to market in 1999. Honda's hybrid system depended on an ultra-thin brushless electric motor that was located on the crankshaft of its internal combustion engine. Here, it acted as a starter motor, an engine balancer, and provided additional power to the wheels. However, the Honda concept had important differences from the Toyota one: unlike the Prius, the Insight was not designed to run on electricity alone, but its electric motor essentially allowed

Honda's Insight had a shape distantly reminiscent of the General Motors EV1. In this case, the electric motor could not power the car on its own; it was purely used to assist the petrol engine.

a smaller and less polluting petrol engine to do the job of a larger one.

Stung into action by these Japanese initiatives, and by the success of the fully-electric Tesla (see next chapter), General Motors in the USA decided to return to the fray. It took time: the new Chevrolet Volt was previewed as a concept in 2007 but did not enter production until 2011. Exports followed a year later, and the car was also sold in other countries as the Ampera, with various GM brand names attached.

The Volt was designed as a Plug-in Hybrid, with a 40-mile range in electric mode. GM reasoned that a distance of 40 miles was the average daily commute in the USA and that the batteries could then be recharged from the mains overnight. The Volt could therefore operate as a pure electric vehicle for most of the time, but it also had a range-extender petrol engine (sourced from European subsidiary Opel) that could recharge the battery through a generator when its voltage fell below a predetermined level. A sophisticated control system made the petrol engine cut in automatically when required, and could

The Chevrolet Volt had a range-extender petrol engine and delivered very acceptable levels of performance but still only 40 miles in pure electric mode. This is the Vauxhall Ampera version of the car that was sold in the UK.

From 2020,
Volvo offered the
XC40 Recharge
family SUV as a
PHEV with a
21-mile electric-
only range, or
a Pure Electric
model with a
range of up to
259 miles.

distribute the power from the range-extender engine to both the generator and the driven wheels, dividing it as appropriate to conditions at the time. A regenerative braking system also contributed to charging the batteries in all operating modes.

The Volt was well received. Its 100mph top speed was a big improvement over the governed 80mph of its EV1 predecessor, and the 0–60mph acceleration time of just over 9 seconds in fully-electric mode was quite acceptable. GM guaranteed the battery pack for eight years or 100,000 miles, which removed another owner concern. However, the obvious savings in fuel costs had to be set against the car's relatively high cost, which had to be amortised over many thousands of miles. As a result, the Volt's real appeal was to buyers who were more interested in its zero-emission credentials than in saving fuel costs.

In mixed use, British magazine *What Car?* said that the UK-spec Vauxhall Ampera actually delivered only 20–30 miles on battery power alone, and that its petrol consumption worked out at between 40 and 60 mpg. It was not enough to convince buyers, and GM decided against making right-hand-drive versions of the second-generation Volt, which was introduced in 2016 with improvements such as a larger range-extender engine, 20 per cent more battery storage capacity, and lighter electric motors.

Nevertheless, these three cars set the scene for the many hybrid models that have since become available. Toyota developed its hybrid technology further and rolled it out progressively across other model-ranges, usually as an alternative to a conventional internal combustion engine. Honda made the second-generation Insight a more conventional-looking family hatchback in 2009, designing the car deliberately to make hybrid technology more affordable to a wide range of buyers. And Chevrolet persevered with the Volt until 2019, dealing on the way with concerns about various safety hazards presented by its new technology and introducing other hybrid models as well.

In 2019, Volvo became the first manufacturer to offer PHEV versions of its entire model range, and it was clear that the industry was committed to hybrids for the immediate future. There was just one dark cloud on the horizon: both the US State of California and the United Kingdom had set a target date of 2035 (since pulled forward to 2030 in the UK) to put a ban on the sales of cars with an internal combustion engine – and that, of course, would include a ban on hybrids as well. For the longer-term future, the pure-electric vehicle would have to be made fully dependable and more affordable.

Graphics show the layout of the Kia Niro PHEV at the 2017 Los Angeles Auto Show. The electric components were mounted low down to optimise use of space and lower the vehicle's centre of gravity.

MODERN TIMES

IN THE FIRST two decades of the twenty-first century, the spread of the electric car was very much aided by incentive schemes set up by the governments of several countries, such as purchasing subsidies and tax allowances. Behind these were attempts to meet emissions targets in accordance with international agreements, and in the second decade of the century some countries set out road maps for phasing out cars that were dependent on internal combustion engines.

By the end of 2019, the pure-electric solution had become roughly twice as successful globally as the hybrid option – although that was not obvious in the West, because more than half of the world's electric cars were in China (and most Chinese models were available only in their own domestic market). The Chinese had caught up very rapidly indeed, and International Energy Agency figures from the start of 2019 showed that the country had 2.58 million battery-electric vehicles, as against 0.97 million in Europe and 0.88 million in the USA. There was also a well-established public charging network.

Most major manufacturers around the world had an electric model in their catalogues after 2010, even if it was only a little-bought option on one of their smaller models. The proliferation of models was of course greatly helped by improvements in technology – most notably lithium-ion battery packs – and it was clear that the electric car was becoming part of the automotive mainstream even if it

OPPOSITE
Rapid Charge points were an essential evolution, offering ever shorter recharging times. This picture of a Nissan Leaf on charge suggests how the familiar 'filling station' might evolve.

Charging plugs generally resemble the nozzle ends of petrol pumps, but plug into a socket on the car. This is a BMW i3 on charge in Germany.

In the Tesla Model S, the batteries were not built into the floor of the car but actually formed the platform on which the car was built.

was still numerically very much inferior to its combustion-engined counterpart.

There was considerable variation among the many models that became available in this period, but three cars in particular deserve comment. From the USA, there was the Tesla Model 3; from Japan, there was the Nissan Leaf; and from Europe there was the BMWi3. In that order, they were also numerically the most successful electric models of their time.

The Tesla Model 3 laid claim to global sales of 500,000 between 2017 and 2020 – a figure which was simply unthinkable for any earlier production electric car. Tesla Motors was established in California in 2003, and from the start, the company adopted business strategies quite unlike those of conventional car manufacturers. These certainly aided customer acceptance of its products, which also incorporated advanced new technology.

Tesla recognised that a major problem in earlier electric-car programmes had been their association with economy motoring, which conflicted with the high cost of the battery technology required. The company's solution was to accept the high cost and to build electric cars for wealthy buyers, allowing these to prove the technology and promote wider public acceptance of electric cars. In the meantime, advances in battery technology and reductions in cost would make lower-priced models feasible, and these would take the Tesla brand into the mainstream car market.

Tesla's first electric car was therefore a luxury two-seat Roadster, based on the chassis of the British-made Lotus Elise sports car and clothed in an attractive new body. The Tesla Roadster entered production in 2008 and was sold exclusively online in another mould-breaking business strategy that dispensed with the intermediary (and cost) of dealers.

It immediately made clear that the traditional objections to electric power could now be overcome. The Roadster offered a range that was calculated by the US

SUVs were popular and fashionable, and the Tesla Model X tapped into that, adding the 'surprise and delight' appeal of gullwing rear doors.

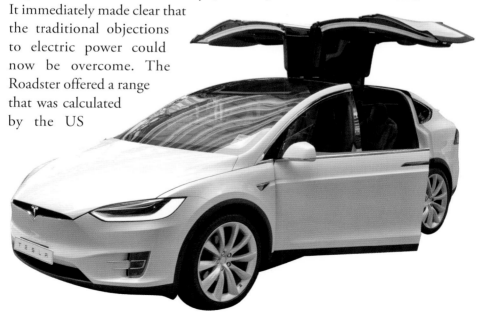

EPA at 244 miles on a single charge; its batteries could be fully recharged from a domestic or other dedicated power supply in about three and a half hours; it had a top speed of 125mph and could accelerate from rest to 60mph in under four seconds.

As a demonstration of technology, this was a huge step forward. It was largely made possible by a lithium-ion battery pack that consisted of multiple cylindrical cells like those used in consumer electrics. This was mounted under the floor to ensure minimum intrusion into passenger and luggage areas and drove through a single electric motor that developed 248bhp in initial guise; it was later improved to 288bhp.

Tesla built 2,500 examples in four years, ending production in 2012 mainly because Lotus wished to end manufacture of its Elise chassis.

The Roadster was followed in 2012 by the Model S, a hatchback luxury saloon, and then by the Model X in 2015, configured as an SUV to suit prevailing tastes. In 2017, the Model 3 was a medium-sized four-door saloon that brought the technology to a more affordable level; this had a typical range of 263 miles (or 220 miles in a low-cost version and 353 miles with an extra-cost Long Range specification). It was an immediate sales success and attracted 325,000 orders globally in the first week after it was announced.

If Tesla's spectacular success led the way for acceptance of electric cars, the Nissan Leaf of 2010 showed an alternative recipe for success. The Nissan strategy was to develop a dedicated structure for the car, but to make it look like an everyday five-door family hatchback. The lithium-ion battery pack was located below the seats and the rear footwell, and powered the front wheels through a 107bhp electric motor. With 93mph and 0–60mph acceleration of under 10 seconds,

The Nissan Leaf was improved over time just like a conventional car, and remained a strong seller. This is a 2018 model.

OPPOSITE Stylish, but deliberately 'safe' as well, the 2010 Nissan Leaf brought reliable battery-electric technology to a yet more affordable level. The decals were not standard; this car was on display at the Tokyo Motor Show in 2009.

The BMW i3 was considered reliable enough to be used by the police in Italy, although the police livery did nothing for the car's deliberately quirky lines.

the car's performance was entirely in keeping with expectations. The first versions had a range of just 73 miles (as calculated by the American EPA) and a full battery charge took seven hours, but Nissan followed a programme of gradual improvements over the car's production life.

The Leaf was more expensive than a comparable car with a conventional combustion engine, but it was essentially unthreatening and proved reliable. Between 2011 and 2014 it was the world's best-selling electric car, being built in the UK and the USA as well as Japan after 2013. The car that stole its crown was the Tesla Model S, but Nissan countered from 2018 with a second-generation Leaf, using the same technology but now with more power and a longer range of 151 miles (or 226 miles with an optional larger battery). By late 2020, global sales of the two generations of Nissan Leaf were approaching 500,000.

Like Nissan, the German BMW company had a strong global reputation which undoubtedly helped when it introduced its first electric production models in 2013. The BMW strategy shared some elements with the one adopted by Tesla, and the first concept car was a large sports coupé that was aimed at wealthy customers. It was displayed at the Frankfurt Motor Show in 2009 with the cumbersome title of BMW Vision EfficientDynamics, but it was actually the forerunner of the electric coupé that would enter production in 2014 as the BMW i8, a deliberately prestigious and expensive model. However, two years after that first concept, BMW announced a second concept, this time sized as a supermini and aimed at a very different group of customers. This went on to become the i3 production model in 2013.

For the i3, BMW used a quirky appearance to give the car a special appeal, and the car was launched as a tall-bodied hatchback with large wheels that gave an almost toy-like appearance. Its powertrain had been designed on the premise that the battery would typically be recharged only every two or three days, and depended on a lithium-ion power pack mounted under the floor with a 170bhp electric motor driving the rear wheels. The typical range on a single charge was 100 miles, but the i3 had three selectable drive modes, which could emphasise performance at the expense of range or vice versa. The pure-electric i3 was also supplemented by a hybrid version,

The success of familiarity… Renault made their electric Zoe model look like any other small hatchback. This is a 2013 model.

which used the 647cc two-cylinder petrol engine from a BMW scooter as a range extender but, in the absence of a multi-range gearbox, could not be used on its own for long-distance work; this extra-cost option increased the range to 150 miles.

Gradual improvements to the i3 followed the pattern established by other electric-car makers, and a second-generation model introduced for 2017 brought an increase in battery capacity of more than 50 per cent. By October 2020, i3 sales had reached 200,000.

By 2020, increasing numbers of pure-electric vehicles were reaching the market and the technology had made great strides. Renault offered smaller cars such as the Twingo and Zoe, the latter with a 245-mile range and a rapid-charge facility that provided 90 miles of range after a 30-minute charge. Higher up the market was new brand Polestar, announced in 2017 as an all-electric sister brand to Volvo in Sweden and a joint venture with Volvo's Chinese owners, Geely. Following the Tesla marketing pattern of introducing an expensive model first, Polestar began with an expensive PHEV luxury coupé called Polestar 1, but in early 2019 came the Polestar 2, a pure battery-electric four-door saloon with 400bhp split between motors driving the front and rear wheels. It was acclaimed for its style, build quality, performance, and a range of 290 miles or more, and, along with the rival Tesla 3, was a clear indication of how far the pure-electric car had now developed.

The cost of an electric car was still considerably greater than that of its conventional equivalent, by as much as 20 or 30 per cent in some cases. Industry experts anticipated that battery costs would fall further and reduce this disparity in prices, but this would depend both on the development of new battery technology and on increases in production allowing prices to be reduced. Until then, car manufacturers simply had to sit it out, building sales of electric vehicles and accepting them as a lower-profit product.

Government subsidies meanwhile continued to drive sales. This was especially so in Norway, where major subsidies had brought electric cars firmly into the mainstream. In 2020, Renault admitted that its strategy depended at least in part on making use of government subsidies, and in France that year zero-emission Renault models outsold diesel versions for the first time (although only just: the figures were 19 per cent and 18 per cent of total Renault sales respectively). However, it was notable that in low-income countries and in those without government subsidies, the take-up rate for electric cars was very low indeed.

Much also depended on the rollout of an infrastructure of electric charging points, without which even the long range of the latest electric vehicles lost some of its value. Some governments supported such schemes; others did not. But the stage was set for the electric car to replace its petrol and diesel alternatives – although it was likely to be many years before the change was complete.

Devastatingly stylish, the Polestar 2 saloon was well received in spite of its relatively elevated price.

FURTHER READING

Anderson, Curtis Darrel, and Anderson, Judy. *Electric and Hybrid Cars: A History.* McFarland, 2005.

Kirsch, David A. *The Electric Vehicle and the Burden of History.* Rutgers University Press, 2000.

Larminie, James, and Lowry, John. *Electric Vehicle Technology Explained.* John Wiley & Sons, second edition, 2012.

PLACES TO VISIT

In addition to the permanent exhibitions listed here, many motor and other museums hold special electric-vehicle exhibitions from time to time.

Grampian Transport Museum, Alford, Aberdeenshire, AB88 8AE. Telephone: 01975 562292.
Website: www.gtm.org.uk The museum has a replica of the electric powertrain made by local inventor Robert Davidson in 1839.

Route 66 Electric Vehicle Museum, 120 W, Andy Devine Ave, Kingman, AZ 86401, USA. Website: www.hevf.org This is the world's only specialist museum for electric vehicles.

Transport Museum Wythall, Chapel Lane, Wythall, Worcestershire, B47 6JA. Telephone: 01564 826471.
Website: www.wythall.org.uk This museum has an important collection of battery-electric light commercial vehicles, built between the 1930s and the early 1980s.

ACKNOWLEDGEMENTS

Alexander Migl/CC-BY-SA-4.0, pages 8 (bottom), 57, 61; Alf van Beem/ Public Domain, page 18 (top); American Motors Public Relations Department/ PhotoQuest/Getty Images, page 25 (bottom); andrewrabbott/CC-BY-SA-4.0, page 17 (top); Arnaud 25/CC-BY-SA-4.0, page 10 (bottom); Bahnfrend/CC-BY-SA-4.0, page 55; Bettmann/Getty Images, page 22; BMW Group Archives, pages 33, 34, 35; Bob1960evens/CC-BY-SA-4.0, page 17 (bottom); Buch-t/CC-BY-SA-3.0-DE, page 19 (top); Bundesarchiv 183-21519-0005 Neue Fahrzeuge der Deutschen Post (DDR), page 19 (bottom); Chanokchon/CC-BY-SA-4.0, page 50; Claus Ableiter/CC-BY-SA-2.0, pages 9, 40 (bottom), 41; Daimler AG, pages 37, 38; dave_7/CC-BY-SA-2.0, pages 21, 48; DeFacto/CC-BY-SA-3.0, page 24 (top); Franz Haag/Public Domain, page 6; FREDERIC J. BROWN/ AFP via Getty Images, page 51; Grey Geezer/Public Domain, page 59; Gzzz/ CC-BY-SA-4.0, page 58; Henrysirhenry/CC-BY-SA-3.0, page 7 (bottom); In Pictures Ltd./Corbis via Getty Images, page 52; Kārlis Dambrāns/CC-BY-2.0, page 54 (top); Courtesy of Land Rover, page 46; Library of Congress/Public Domain, pages 12–13; Lsd lysergid/CC-BY-SA-4.0, page 32; MattiBlume/ CC-BY-SA-4.0, page 39; Mike Gould, page 45; NARA/Public Domain, page 28; NMM Heritage Images/Getty Images, page 20; Oleg Alexandrov/CC-BY-SA-3.0, page 54 (bottom); Olli1800/CC-BY-SA-3.0, page 27; Ominae/CC-BY-SA-4.0, page 36; Pierre Vauthey/Sygma/Sygma via Getty Images, page 42 (top); Plug In America/CC-BY-SA-2.0, page 43 (bottom); Public Domain, pages 7 (top), 10 (top), 11, 16; Rainerhaufe/CC-BY-SA-4.0, page 30; Right Brain Photography (Rick Rowen)/CC-BY-SA-2.0, pages 42, 43 (top); Skartsis/CC-BY-SA-3.0, page 24 (bottom); SSPL/Getty Images, page 29; Stock Montage/Getty Images, page 4; Tennen-Gas/CC-BY-SA-3.0, page 56 (bottom); Thesupermat/ CC-BY-SA-4.0, page 18 (bottom); treeday77/CC-BY-2.0, page 49; Txemari/ Public Domain, page 40 (top); Vauxford/CC-BY-SA-4.0, pages 8 (top), 25 (top), 56 (top); Wally McNamee/CORBIS/Corbis via Getty Images, page 26.

INDEX